ISBN: 9798378609918 (Paperback)
ISBN: 9798377771715 (Spanish edition)

Cover design by: Fernando Zambrano
Printed in the United States of America

ARTIFICIAL INTELLIGENCE IN ENGINEERING

Discovering Unknown Patterns and Solutions

Fernando Zambrano P.E.

Kindle Direct Publishing

To those whose passion and dedication have enabled the marriage of artificial intelligence and engineering, thank you for leading the transformation of our world.

CONTENTS

PROLOGUE

It is an honor to write the prologue for this book that explores the revolutionary application of Artificial Intelligence (AI) in engineering. From definition to applications, history, types of AI, and challenges in its implementation, this book provides a comprehensive and detailed view of how AI can enhance current methods of computer simulation and mathematical calculation.

Artificial Intelligence is a field in constant evolution and has been growing and accelerating in recent years. It's incredible to see how AI has been able to revolutionize the way we live and work, and it's equally exciting to see how it can be applied in engineering. From pattern recognition to building bridges and dams, AI can help improve engineering projects' efficiency, safety, and quality.

The book presents a wide range of machine learning techniques and algorithms, such as linear regression, decision trees, and neural networks, and their application in engineering. Furthermore, it compares traditional simulation and mathematical calculation models with AI-based models, offering a comprehensive view of the improvements that can be achieved by implementing AI in engineering.

In Chapter 7, the book provides conclusions and recommendations on the application of AI in engineering, showcasing the importance of this technology in the future. The future prospects of AI in engineering are fascinating and will undoubtedly open new avenues for innovation and progress in this field.

It is important to note that the author of this book used in part GPT-3.5, OpenAI's large-scale language generation model, as a tool for creating the initial content. However, it should be noted that the author reviewed, edited, and adjusted the content to his style, reflecting his unique ideas and perspectives. As a result, this book is presented as a derivative work with an authentic and individual vision in which the author assumes full responsibility for the final content of this publication.

In conclusion, this book is an excellent guide for anyone interested in the application of AI in engineering and how it can enhance current simulation and mathematical calculation methods. Furthermore, this book inspires readers to explore this field further and understand its potential to revolutionize the way engineering is done.

Without further ado, I hope you enjoy reading and learning a lot about AI in engineering.

INTRODUCTION TO ARTIFICIAL INTELLIGENCE (AI)

Technology has transformed the world in a way never seen before, and artificial intelligence is a fundamental part of this revolution. From task automation to process improvement and decision making, artificial intelligence is revolutionizing the way we live and work. In the field of engineering, artificial intelligence is discovering unknown patterns and solutions that are helping to improve simulation and mathematical calculation methods.

1.1 Definition of Artificial Intelligence (AI)

AI is a branch of computer science that develops algorithms and techniques that allow machines to simulate human intelligence. It's a set of methodologies that will enable computers to learn, reason, and make decisions autonomously.

AI is one of the most exciting and promising areas in the field of technology. With its ability to learn, reason, and make decisions, AI has revolutionized how businesses and individuals interact with the digital world.

1.2   History of Artificial Intelligence (AI)

AI has been a topic of interest and study since antiquity, with concepts related to robotics and AI appearing in Greek and Chinese literature. The idea of creating machines capable of imitating human intelligence has fascinated philosophers, scientists, and mathematicians throughout history. Since then, humanity has sought to understand how intelligence works and how it can be replicated in machines.

However, the modern development of AI began in 1956 when a group of scientists and mathematicians gathered at Dartmouth College to discuss the possibility of creating intelligent machines. This meeting is considered the beginning of the modern field of AI and marked a new era of research and development in AI technology. During the following years, researchers focused on developing algorithms that could imitate human intelligence and create systems capable of learning and reasoning.

As technology advanced and computational resources became more accessible, AI began to develop more effectively. As a result, new techniques and algorithms have been developed, including machine learning-based AI and knowledge-based AI, allowing machines to learn and reason more effectively.

Additionally, AI has significantly impacted society, the economy, and daily life. It is used in various fields, including medicine, finance, energy, and security, and is expected to continue to evolve and improve in the future. However, AI has also generated a series of ethical and social challenges, including privacy, security, and responsibility. Therefore, it is important that the scientific community and society as a whole continue to discuss and address these challenges as AI continues to evolve and develop.

In conclusion, AI is a constantly evolving and developing field with a rich history and a promising future. Since its inception in 1956, AI has experienced an impressive evolution and significantly impacted society and the economy. As technology advances, AI will continue to play an essential role in daily life and in solving complex problems in various fields. Therefore, the study and understanding of AI is fundamental in maximizing its potential and addressing the ethical and social challenges that arise with its use. In this book, we will explore how the application of AI in engineering can help improve current methods of computer simulation and mathematical calculation.

1.3 Types of Artificial Intelligence (AI)

In recent years, AI has experienced rapid growth and significantly impacted many industries, including engineering.

In AI, there are three types of artificial intelligence: weak, strong, and general. Each class focuses on different aspects of human intelligence and has other goals and applications.

1.3.1 Weak Artificial Intelligence

Weak Artificial Intelligence, also known as specialized or limited AI, focuses on developing systems that are experts in a specific task, such as image recognition, natural language processing, or rule-based decision-making.

An image recognition system can be considered an example of weak AI, as it focuses on identifying specific objects or features in images but does not need to understand the context or broader significance of the pictures.

Aside from image recognition, weak AI can be applied in various tasks such as text classification, fraud detection, rule-based decision-making, control of automated systems, and process optimization. Weak AI is a valuable technology in the business world, as it can help automate repetitive tasks, improve efficiency, and reduce errors.

However, despite its many benefits, weak AI has some limitations. For example, its ability to solve problems is limited, and its effectiveness largely depends on the quality and quantity of data provided. Therefore, it is vital to remember that results obtained from weak AI may be inaccurate or imprecise, especially if the data used is incomplete or incorrect.

In addition, weak AI can also be vulnerable to attacks or intrusions, which can be harmful to businesses and users. For example, an attacker can manipulate the data used by a weak AI system to obtain false or unexpected results. For this reason, it is essential to implement security measures and constantly monitor vulnerable AI systems to prevent and detect possible vulnerabilities.

Despite these challenges, weak AI is a valuable and widely used technology in various fields and applications, and its use and development continue to evolve. For example, in the field of engineering, weak AI can be used to improve the efficiency of simulation and mathematical calculation processes on a computer. For example, it can be used to identify patterns in data, automate decision-making and optimize complex processes.

In summary, weak AI is a limited type of AI used to solve specific tasks such as image recognition, natural language processing, or rule-based decision-making. It is a valuable technology that can help improve efficiency and reduce errors in various applications, including engineering. However, it is also essential to consider its limitations and take measures to prevent potential vulnerabilities and ensure accurate results.

1.3.2   Strong Artificial Intelligence

Strong Artificial Intelligence, also known as autonomous AI, focuses on developing systems that can perform any intellectual task that a human being can perform. This type of AI seeks to imitate human intelligence in all aspects, including language understanding, learning, and problem-solving. Although strong AI has not yet been fully developed, it is a very active area of research and has great potential to revolutionize many fields, including engineering. Some examples of this AI are:

A. Machine Learning

Strong AI is based on machine learning, a process by which computers can improve their performance without being specifically programmed to perform a particular task. Instead, this process is based on data collection, pattern recognition, and decision-making based on these patterns.

B. Neural Networks

Neural networks are a vital technique in strong AI. These networks imitate the structure and behavior of neural networks in the human brain, allowing computers to learn and make decisions based on available information.

C. Supervised and Unsupervised Learning Algorithms

Supervised and unsupervised learning algorithms are critical techniques in machine learning and strong AI. Supervised learning algorithms are based on collecting labeled data and learning based on patterns. Unsupervised learning algorithms are based on pattern recognition without the help of labeled data.

D. Applications of Strong AI

Strong AI has a wide variety of applications, including process automation, improving energy efficiency, detecting failures in industrial systems, improving safety in transportation systems, and improving the quality of life of people with disabilities.

E. Challenges in Developing Strong AI

Despite its potential, the development of strong AI presents a series of challenges, including data privacy and security, ethics and responsibility in decision-making based on AI, and the need for adequate regulations to ensure the responsible use of technology, among others.

F. Future Perspectives of Strong AI

Strong AI has enormous potential to transform engineering and other areas of society further. In the future, more advanced technologies will be developed that will allow computers to imitate human intelligence even more and improve the efficiency and effectiveness of simulation and mathematical computation processes on computers. Additionally, it is expected that strong AI will positively impact the economy, creating new jobs and improving productivity. However, it is vital to consider the ethical and legal challenges that may arise with the advancement of technology and take appropriate measures to ensure its responsible use. In summary, strong AI has limitless potential, and its application in engineering and other fields can help significantly improve human life.

1.3.3   General Artificial Intelligence

General Artificial Intelligence, also known as human-level AI, refers to artificial intelligence capable of performing any task a human can perform. This form of AI is still an object of study and development. Still, it is expected to be capable of performing complex tasks, learning, and improving over time, and having deep knowledge and understanding of the world.

Below, we will describe some current and future examples of the applications of AI in engineering.

1. Machinery fault prediction: General AI can be used to analyze large amounts of data and detect patterns that may indicate a machinery failure. This can help prevent accidents and improve production efficiency.

2. Product design and optimization: General AI can be used to simulate and evaluate different design options and determine the best option. It can also help optimize the production of a product, reducing costs and improving quality.

3. Industrial process monitoring and control: General AI can be used to monitor and control industrial processes in real-time. This can help improve efficiency and prevent errors and waste.

4. Demand prediction: General AI can be used to analyze historical demand data and predict future demand. This can help improve planning and production optimization.

5. Scientific research: General AI can be used to analyze large amounts of scientific data and help identify patterns and unknown solutions. This can be especially useful in researching new materials and technologies.

6. Development of advanced materials: General AI will be applied to develop more durable, efficient, and resilient materials for building structures and buildings.

7. Improvement in urban planning and sustainable development: General AI will

be used to create more efficient and sustainable urban planning, using data analysis and advanced technology to improve people's quality of life.

8. Energy optimization in buildings and homes: General AI will be used to optimize energy use in buildings and homes, improving energy efficiency and reducing costs and environmental impact.

9. Improved transportation efficiency: General AI will be applied to optimize routes, improve safety, and reduce waiting times.

10. Natural disaster prediction: General AI will be used to predict natural disasters such as earthquakes, hurricanes, and floods, allowing people to be better prepared and avoid damage to property and human life.

11. Advanced medical diagnosis: General AI will be used for more precise and efficient medical diagnoses, improving patient care and treatment.

12. Improved security at airports and ports: General AI will be used to enhance the security at airports and ports, using facial recognition technology and data analysis to detect threats and prevent attacks.

13. Improvement in agricultural efficiency: General AI will be used to improve efficiency in agriculture, using data analysis and advanced technology to optimize production, enhance the quality of food and reduce waste.

14. Advancements in Robotics: General AI will play a significant role in developing advanced robotics, making robots more intelligent, autonomous, and able to perform complex tasks.

15. Customer Service: General AI will be used to provide advanced customer service by using natural language processing and machine learning algorithms to understand and respond to customer inquiries.

16. Fraud Detection: General AI will be used to detect fraud by analyzing large amounts of data and identifying patterns that indicate suspicious activity. This will help to prevent financial losses and improve security.

17. Enhancements in Cybersecurity: General AI will be used to enhance cybersecurity by detecting and preventing cyber-attacks, detecting malicious activities, and analyzing large amounts of data to

identify potential vulnerabilities.

18. Improving Supply Chain Management: General AI will improve supply chain management by optimizing processes, predicting demand, and reducing costs.

19. Predictive Maintenance: General AI will be used to predict when equipment will fail, allowing preventative maintenance to be performed before the equipment stops working. This will reduce downtime and improve efficiency.

20. Improved Natural Language Processing: General AI will improve natural language processing by developing advanced algorithms that can understand and generate human-like speech and writing.

These are just a few examples of how general AI can improve the application of data analysis in engineering and help discover unknown patterns and solutions. AI in engineering is a constantly evolving field, and its application opens the door to new solutions and possibilities that have not yet been discovered. Data analysis in engineering is a valuable tool for improving processes and achieving greater efficiency, and AI is a perfect complement to this task. With its ability to identify patterns and unknown solutions, AI in engineering is an indispensable tool for the future of engineering and the resolution of complex problems.

In summary, AI in engineering offers a series of unknown solutions and benefits, with a focus on improving computer simulation and mathematical calculation. The application of AI in engineering can help in the prevention and solution of problems, the optimization of processes, integration with existing technologies, customization of solutions, improvement in collaboration, ease of use, and cost reduction. Additionally, it allows for greater safety in the construction and design of projects, as well as greater efficiency and quality in project execution.

This chapter has only touched the surface of the applications of AI in engineering. By the end of this book, the reader will have a clear and deep understanding of how AI can be used in engineering to improve current methods of computer simulation and mathematical calculation, and how this technology is revolutionizing the industry. Are you ready to discover how AI can help you solve more complex problems and improve your processes? Keep reading to find out!

ARTIFICIAL INTELLIGENCE (AI) IN ENGINEERING

A I is a field in constant evolution and increasingly significantly impacts many aspects of daily life, including engineering. The ability of AI to learn, analyze, and make real-time decisions is revolutionizing how engineers approach challenges and solve problems in their field. From simulation and modeling of systems to process optimization and automation, AI is transforming how engineering is practiced. In this chapter, we will explore AI's current and future applications in engineering, including how AI is being used to improve efficiency and accuracy in solving complex problems in the real world.

2.1   Applications of Artificial Intelligence in Engineering

Civil Engineering: This is a discipline that deals

with the design and construction of structures, bridges, roads, ports, and dams, among others. It is one of the oldest and most varied disciplines within engineering and is involved in creating and improving urban infrastructure.

AI is transforming civil engineering in many ways. For example, machine learning algorithms can analyze large amounts of geotechnical data and predict potential problems in the construction process. Additionally, AI can optimize the design and construction of structures, resulting in more efficient and sustainable construction. Deep learning techniques are also being used to predict bridges and road failures and develop preventive solutions.

Also, AI can be used to perform real-time structure inspections, allowing for the early detection of problems and the prevention of collapses. Additionally, AI can be used to predict the behavior of structures in adverse conditions, such as earthquakes and floods, resulting in safer structures.

In summary, civil engineering and AI are evolving together, allowing for greater efficiency and accuracy in the design and construction of structures. AI is transforming civil engineering, creating a safer and more sustainable infrastructure.

Case Study: Construction project management

using AI.

Objective: Optimize scheduling and resource allocation in a skyscraper construction project.

Step 1: Gathering project data and parameters.
- Data is collected on project tasks and activities, as well as available resources and their associated costs.
- Project parameters such as total duration, time constraints, and cost targets are defined.

Step 2: Generation of the project schedule using AI
- An AI algorithm generates an optimal project schedule, considering the dependencies between tasks, required resources, and time constraints.
- The algorithm analyzes historical data from similar projects and looks for patterns that indicate the best sequence of tasks and resource allocation to minimize project duration and associated costs.

Step 3: Resource Allocation Optimization
- Using machine learning techniques, resource allocation optimization is performed to maximize efficiency and minimize costs.
- The algorithm analyzes real-time data, such as task progress and resource availability, and adjusts the allocation to avoid delays and optimize productivity.

Result:
- Using AI, overall project duration was reduced by 10% compared to traditional scheduling.
- A 5% decrease in total project costs was observed

due to optimized resource allocation.

This case study shows how the application of AI in civil engineering can deliver significant results in both efficiency and cost. Furthermore, engineers can leverage these techniques and tools to improve the quality and performance of their projects while maintaining a focus on safety and sustainability.

Electrical Engineering: This is a branch of engineering that focuses on the design, development, and maintenance of electrical and electronic systems and equipment. AI is increasingly being integrated into this field to improve the efficiency and effectiveness of electrical systems.

One of the most common applications of AI in electrical engineering is monitoring and controlling electrical systems. For example, AI monitors electrical demand, identifies consumption patterns, and optimizes energy generation to meet demand more efficiently.

Additionally, AI is used in diagnosing and solving problems in electrical systems. AI algorithms can analyze fault and diagnostic data to identify patterns and predict future issues, allowing for a faster and more efficient solution.

AI technologies are also used in planning and designing electrical systems. For example, AI algorithms can analyze energy consumption patterns and simulate different design scenarios to identify the most efficient and cost-effective option.

In conclusion, the integration of AI in electrical engineering is transforming the way electrical systems are monitored, controlled, and designed, allowing for greater efficiency and a faster solution to problems.

Mechanical Engineering: Mechanical engineering is a branch of engineering that deals with developing and improving mechanical systems, components, and devices for a wide range of applications. This discipline includes manufacturing, designing, developing, and maintaining machines, tools, vehicles, and other mechanical systems.

AI is a technology that is changing the world, and mechanical engineering is no exception. The application of AI in mechanical engineering can help improve the design and manufacturing process of mechanical systems, reducing costs and increasing efficiency.

AI algorithms can be used to optimize manufacturing processes, reducing the time and cost of production while improving quality. Additionally, AI can be used to develop predictive maintenance solutions, allowing for the proactive detection and resolution of potential problems before they occur.
In conclusion, the integration of AI in mechanical engineering is transforming mechanical systems'

design and manufacturing process, resulting in reduced costs and improved efficiency.

Case Study: Optimization of mechanical component design using AI.

Goal: Use artificial intelligence to optimize the design of mechanical components, improving their performance and reducing development time.

Step 1: Data collection
- Design and performance data is collected on existing mechanical components, including information on geometry, materials, and mechanical properties.
- Data is collected on performance requirements, such as maximum load, fatigue strength and required stiffness.

Step 2: Data analysis and preprocessing
- The collected data is analyzed to identify relationships between design variables and component performance.
- Data preprocessing techniques are used to normalize the data and eliminate outliers, ensuring greater accuracy in the results.

Step 3: AI model development
- A machine learning algorithm, such as a genetic algorithm or particle swarm-based optimization algorithm, is used to develop a model to find the optimal mechanical component design.
- An objective function representing the desired

performance is defined, and design constraints, such as manufacturability and economic feasibility, are established.

Step 4: Component design optimization
- Using the AI model, iterations are performed to find the optimal mechanical component design, considering different combinations of design variables.
- The performance of each proposed design is evaluated, and the design that meets the performance requirements and the established constraints is selected.

Aerospace Engineering: In aerospace engineering, AI is used to improve aircraft construction safety and efficiency. For example, AI algorithms are being developed to optimize the distribution of materials and resources in aircraft construction and to predict failures in aerospace components. Deep learning techniques are also being used to improve aircraft navigation efficiency and develop preventative solutions in case of failures. In addition, AI is being used to improve air traffic control safety and optimize flight route planning.

Hydraulic Engineering: Hydraulic engineering is a branch of civil engineering that deals with planning, designing, constructing, and managing systems and structures that allow the handling, controlling, and distributing of water in a territory. AI can significantly help in this field, allowing for improved efficiency and accuracy in water management processes.

For example, AI can help to model hydraulic systems with greater accuracy, allowing for better prediction of their operation and potential failures. In addition, AI technology can also help optimize water distribution and control processes using machine learning algorithms and predictive models.

Furthermore, AI can greatly help detect and prevent leaks in hydraulic systems using sensors and advanced leak detection technologies.

In summary, AI can bring many solutions to hydraulic engineering, allowing for improved efficiency, accuracy, and safety in water control and distribution systems.

Sanitary Engineering: Sanitary engineering is a branch of engineering responsible for ensuring the provision of potable water and wastewater treatment to ensure the public health and the environment. The application of AI in sanitary engineering can improve the optimization process in the distribution and treatment of wastewater, providing more efficient and cost-saving solutions.

AI can be used in sanitary engineering to develop digital representations and predictions of water quality, allowing for more effective management of potable and wastewater networks. In addition, it can also be used to optimize wastewater treatment processes, reducing time and costs in the implementation of solutions.

Another application of AI in sanitary engineering is in detecting leaks in potable water networks, allowing for quick identification and repair, resulting in a significant reduction in costs and improved efficiency in the management of water resources.

In summary, the application of AI in sanitary engineering can improve current processes and provide more efficient and effective solutions in managing water resources. However, it is important to note that this technology must be used responsibly and ethically to ensure public health and the environment.

Process Engineering in the Oil Industry: AI is also revolutionizing process engineering in the oil industry. For example, AI algorithms are being used to improve production processes in the oil industry by predicting equipment failures, optimizing production processes, and improving supply chain management.

In addition, AI can also be used to improve exploration and production processes in the oil industry by predicting the location of new oil deposits and optimizing production processes.

Furthermore, AI can be used to improve the safety of oil industry processes by predicting potential accidents and taking preventive measures.

Systems and Control Engineering in the Oil and Gas Industry: Systems and control engineering is a fundamental discipline in the oil and gas industry, as it is responsible for ensuring the efficient operation of systems and equipment used in oil and gas production and processing. This includes process control systems, equipment monitoring, and automation and safety systems.

The application of AI in systems and control engineering in the oil and gas industry is crucial to improve the efficiency and safety of processes. For example, machine learning algorithms can be used to constantly monitor equipment and detect problems in real-time. This allows engineers to take preventive measures to avoid possible failures or accidents.

In addition, AI is also used to optimize production and processing processes. For example, optimization algorithms can be used to maximize equipment's energy efficiency, resulting in reduced production costs.

In conclusion, systems and control engineering is a crucial discipline in the oil and gas industry, and the application of AI is enabling the improvement of the efficiency and safety of processes and the optimization of oil and gas production and processing.

Road Engineering and AI: Road engineering is a discipline that deals with the design, construction, and maintenance of roads and other road infrastructures. This discipline is crucial for transporting people and goods, as it allows for the efficient and safe flow of vehicles and pedestrians.

Road engineering involves many factors, such as road geometry, construction materials, signage, drainage, lighting, safety, and accessibility. Each of these factors is critical in ensuring the road meets the required safety and efficiency standards.

AI is transforming road engineering in several ways. For example, machine learning algorithms are being used to optimize road design, resulting in greater efficiency in terms of time and cost. In addition, sensor and tracking technologies are being used to continuously monitor traffic and road infrastructure, allowing for rapid detection and resolution of problems.

Geographical information systems are also being used to create detailed maps of road infrastructure, allowing engineers and planners to visualize and analyze information in real time. This allows for better road planning and construction, as well as better resource management and cost reduction.

In conclusion, AI is transforming road engineering, allowing for greater efficiency in designing, constructing, and maintaining roads and other road infrastructures. This results in increased safety and efficiency in the transportation of people and goods.

Ports Engineering and AI: Ports engineering is a branch of civil engineering responsible for planning, designing, constructing, and operating marine and river ports. AI is revolutionizing this field by allowing for more accurate and detailed analysis of weather conditions, port operation modeling and simulation, and process optimization. The AI technology enables the integration of weather, geotechnical, and logistics data to predict and mitigate potential risks in port operations, such as storms and extreme tides. Additionally, using drones and sensors for real-time data collection and 3D modeling allows for better pier location planning, space optimization, and efficient maritime traffic management.

Regarding logistics, AI allows for optimization in cargo and discharge operations management, resource allocation planning, and supply chain optimization. AI can also be used to monitor and optimize security and efficiency in port management, resulting in improved quality and speed in product delivery.

In conclusion, AI is transforming port engineering into a more precise and efficient tool for the planning, constructing, and operating of marine and river ports.

Systems Engineering and Information and AI: Computer systems engineering is responsible for designing, constructing, and maintaining information and information technology systems. AI is transforming this field by allowing for process automation, decision-making, and problem-solving in a more efficient and accurate manner.

One of the most significant advancements in computer systems engineering is automating tasks by implementing software robots and machine learning systems. These systems can perform repetitive tasks such as data entry and system monitoring more quickly and accurately than humans, improving efficiency and productivity.

Another significant advancement is the use of AI for real-time decision-making. For example, AI systems can analyze large amounts of data to predict and mitigate technical problems, improve resource management efficiency, and optimize system security.

Furthermore, AI also allows for improvement in problem-solving in computer systems engineering. Machine learning systems can analyze large amounts of data to identify patterns and solutions to technical problems, resulting in a faster and more accurate resolution of the issues.

In conclusion, AI is transforming computer systems engineering by allowing for process automation, decision-making, and problem-solving more efficiently and accurately. This results in improved efficiency, productivity, and security in handling information systems.

Geotechnics and AI: Geotechnics is a branch of civil engineering that focuses on studying soil and rock materials and their behavior under different loads and environmental conditions. It is used to design and construct structures such as buildings, bridges, and dams and to evaluate their stability and safety.

AI is transforming geotechnics by allowing for more accurate and efficient soil and rock data analysis. Integrating AI algorithms and sensors into geotechnical studies enables the analysis of vast amounts of data to identify patterns and relationships that would be difficult to detect through manual methods. This results in improved accuracy and efficiency in geotechnical assessments and reduced risk in geotechnical projects.

AI is also being used to improve the efficiency and accuracy of geotechnical modeling. Using machine learning algorithms, geotechnical models can be optimized based on real-world data, resulting in improved predictions and soil and rock behavior assessments.

In conclusion, AI is transforming geotechnics by allowing for more accurate and efficient soil and rock data analysis and improving efficiency and accuracy in geotechnical modeling. This results in reduced risk in geotechnical projects and enhanced safety in the design and construction of structures.

Hydrology and AI: Hydrology is the branch of engineering that focuses on studying water resources, including the amount, distribution, and quality of water. Hydrology is essential for water resources planning and management, including the construction of dams, canals, aquifers, and water supply systems, among others.

AI plays a significant role in hydrology by allowing the detection of unknown patterns and solutions in water resources management. For example, machine learning algorithms can be used to analyze large amounts of hydrological data and predict trends in water availability and quality. Additionally, AI can be used to optimize the planning and design of hydrological projects, resulting in a more efficient and sustainable management of water resources. AI can also help monitor water quality and early detection of water-related contamination events, allowing for a faster and more efficient response in critical situations.

In summary, hydrology and AI have a significant relationship, as AI can help improve water resources management, resulting in more efficient and sustainable management of them.

Applications of AI in Environmental Engineering: AI is also significantly impacting environmental engineering, allowing engineers and environmental scientists to solve complex problems and improve the management of natural resources. In this chapter, we will explore the applications of AI in environmental engineering, including water resources management, ecological monitoring, and waste management.

- Water resources management.

Efficient water resource management is critical for environmental and economic sustainability. AI is being used to improve water resources management, from optimizing water distribution to identifying sources of contamination. For example, AI systems are being developed to predict water availability and maximize water resource management in rural and urban communities. Deep learning techniques are also being used to identify sources of contamination in water systems and to develop preventive solutions.

- Environmental monitoring.

Continuous environmental monitoring is essential to understand and manage human ecological impact. AI is being used to improve environmental monitoring, from identifying patterns in ecological data to identifying long-term trends. For example, AI systems are being developed to predict the weather and to optimize air and water quality monitoring. Deep learning techniques are also used to identify environmental data patterns and develop preventive solutions.

- Waste management.

Efficient waste management is essential for environmental and economic sustainability. AI is being used to improve waste management, from optimizing toxic waste management to identifying sustainable solutions for waste disposal. For example, AI systems are being developed to optimize waste sorting and identify the most environmentally friendly waste disposal methods. Deep learning techniques are also used to identify waste data patterns and develop preventive solutions.

In summary, AI significantly impacts environmental engineering, allowing ecological engineers and scientists to solve complex problems and improve the management of natural resources. Furthermore, AI in environmental engineering will enable professionals to apply more efficient, effective, and sustainable solutions to environmental problems.

Applications of AI in Bioengineering: AI is being successfully used in bioengineering to solve complex problems in various areas, including biotechnology, genetics, biomedicine, and many more.

Biotechnology
In biotechnology, AI is being used to improve the design and production of biotechnology products. For example, deep learning techniques are being used to optimize the production of recombinant proteins and to identify new bioactive compounds.

Genetics
In genetics, AI is used to improve genetic disease identification and develop therapeutic solutions. For example, deep learning techniques are used to analyze large amounts of genomic data and identify mutations responsible for genetic diseases.

Biomedicine
In biomedicine, AI is being used to improve disease diagnosis and treatment. For example, deep learning techniques are used to analyze medical images and identify disease patterns. AI systems are also being used to personalize disease treatment, considering each patient's individual genetic and clinical characteristics.

Furthermore, AI is helping to revolutionize biomedical research, allowing researchers to identify new therapeutic targets and develop new treatments more quickly and effectively.

In summary, AI significantly impacts bioengineering, allowing researchers to solve complex problems and improve people's lives in various ways. It is a field in constant evolution and is expected to play an increasingly important role in bioengineering in the future.

BIM Technology in AI: Building Information Modeling (BIM) technology is a methodology for designing and planning buildings that allows the creation of complete and detailed 3D models. These models include information about the structure, materials, facilities, and all relevant aspects of the building. As a result, BIM technology allows for a more complete and detailed project vision, which translates into greater efficiency and effectiveness in decision-making.

AI is transforming BIM technology in several ways. On the one hand, machine learning algorithms are being used to improve the accuracy and realism of 3D models. This is achieved thanks to AI's ability to learn from data and improve the representation of the model's elements.

Furthermore, AI is also allowing real-time optimization of models. This is achieved through AI's ability to continuously analyze and improve models, allowing architects and engineers to make decisions more efficiently.

Finally, AI allows the creation of real-time integrated models. These models combine information from different sources, allowing for a complete and real-time view of the building's components, systems, and data.

AI is revolutionizing BIM technology, allowing for greater accuracy and realism in models, real-time optimization, and integrated model creation. This results in greater efficiency and effectiveness in building design and planning decision-making.

Other Engineering Applications

In addition to the previously mentioned applications, AI is being used in various other engineering fields. For example, AI techniques are being used to improve efficiency in product production and processing in the industry, waste management efficiency, and safety in mining.

- In civil engineering, AI is used for construction project optimization and structure monitoring. In mechanical engineering, it is used for process simulation and modeling and in the control and optimization of systems. In electrical engineering, AI optimizes and controls electrical systems and identifies energy consumption data patterns.

- In aerospace engineering, AI is applied in flight simulation, modeling, and monitoring and controlling aerospace systems. In hydraulic engineering, AI optimizes and maintains water systems and predicts and prevents hydraulic disasters. Finally, in sanitary engineering, AI is employed to optimize and control water treatment and sewage systems.

- In road engineering, AI is applied in traffic optimization, pattern identification, and accident prevention. In process engineering, AI is used to optimize and control industrial processes. Finally, in control engineering, AI optimizes and controls production systems and identifies production data patterns.

- In aerospace engineering, AI is applied in flight simulation and modeling, as well as in monitoring and controlling aerospace systems. In hydraulic engineering, AI is used in the optimization and control of water systems, as well as in the prediction and prevention of hydraulic disasters. In sanitary engineering, AI is employed in the optimization and control of water treatment and sewage systems.

- In road engineering, AI is applied in traffic optimization, pattern identification, and accident prevention. In process engineering, AI is used in the optimization and control of industrial processes. In control engineering, AI is employed in the optimization and control of production systems and in the identification of production data patterns.

- In port engineering, AI is applied in operation optimization, accident prevention, and pattern identification in operation data. In system engineering, AI is used to optimize and control complex systems and identify system data patterns. Finally, in geotechnical engineering, AI is employed in project optimization and in the identification of geotechnical data patterns.

- Hydrology is also being transformed by AI, especially in the field of hydrological event prediction and monitoring. Applying AI algorithms in hydrological models allows for better water body dynamics and flood prediction estimation. AI is also used in water resource management and in the optimization of water supply systems.

In conclusion, AI is becoming a fundamental tool in the innovation and improvement of engineering. From modeling and simulation to process optimization, AI transforms how engineering approaches complex problems and finds efficient solutions.

In addition to its application in engineering, AI has also found an important role in other areas of exact sciences and mathematics. Here are some additional AI applications worth mentioning:

Financial analysis: AI makes financial predictions and analyses to help investors make informed decisions.
Bioinformatics: AI analyzes large amounts of genomic and proteomic data and identifies patterns and relationships among them.

Physics: AI is used in physics to model and simulate complex physical phenomena, including simulating collisions in space and predicting earthquakes and tsunamis.

Mathematics: AI is used in mathematics to solve complex mathematical equations and problems, including number theory and algebraic geometry.

Astronomy: AI is used in astronomy to analyze large amounts of astronomical data and identify patterns and relationships among them, such as identifying exoplanets and predicting solar eclipses.

In astrophysics, AI analyzes large amounts of data obtained from telescopes, allowing astronomers to discover patterns and trends in celestial activity.

In quantum physics, AI is being used to develop algorithms that can simulate and predict the behavior of quantum systems.

In molecular biology, AI analyzes large amounts of genomic and proteomic data, allowing scientists to identify patterns and trends in cellular activity.

Computational chemistry uses AI to simulate and predict chemical reactivity, which can be helpful for the design of new materials and drugs.

In conclusion, AI is a technology that has transformed many areas of exact sciences and mathematics and will continue to be a valuable tool for solving complex problems and improving decision-making in various fields. This book is a guide to discovering how AI is changing engineering and how it can be applied in different areas to improve processes and find unknown solutions.

2.2 Improvements in Current Simulation and Mathematical Calculation Methods

At this point, we will discuss in detail how AI can improve current simulation and mathematical calculation methods in engineering.

In the field of engineering, the application of AI in current methods of simulation and mathematical calculation can have a significant impact in terms of efficiency and accuracy. However, the complexity and amount of data involved in many engineering projects can limit current simulation and mathematical calculation methods. Still, AI can overcome these limitations by using machine learning techniques and optimization algorithms to solve complex problems more efficiently and accurately.

Furthermore, AI can be used to model complex and dynamic systems, allowing for a deeper understanding of the systems and a more accurate evaluation of their performance. The automation of error identification and elimination in mathematical models through AI also results in greater accuracy of the results. The continuous learning and improvement over time of AI lead to improved system efficiency and effectiveness, resulting in improved accuracy and usefulness of decision-making.

For example, the use of AI-based optimization algorithms, such as the genetic algorithm (an optimization technique based on the theory of natural evolution and genetics. It is used to solve complex problems and find the best solution for a given situation), can significantly improve the efficiency of simulation and mathematical calculation in terms of time and resources.

The use of AI also allows for the automation of error identification and elimination in mathematical models, resulting in greater accuracy of the results. Furthermore, AI's ability to learn from historical information and improve over time also enhances the efficiency and effectiveness of the system, as it enables better prediction and more accurate decision-making.

With the use of AI in engineering, it is possible to perform large-scale data analysis efficiently and accurately, significantly improving decision-making and identifying unknown patterns and solutions. Additionally, AI allows for continuous optimization and improvement in system efficiency and effectiveness, improving the quality and effectiveness of engineering projects.

In conclusion, the integration of AI into current methods of simulation and mathematical calculation in engineering results in a more efficient, accurate, and effective solution to the complex problems faced in industry. The continuous learning capability of AI leads to a constant optimization in the efficiency and accuracy of these methods, resulting in a steady improvement in the understanding of systems and more accurate and adequate decision-making in engineering.

Functioning of Major AI Algorithms Applied to Structural Engineering: Structural engineering is a branch of civil engineering that is responsible for analyzing and designing structures to ensure their safety and stability. Artificial Intelligence is an evolving discipline, and research in this field has created increasingly complex and sophisticated algorithms.

Algorithms are a sequence of defined and repetitive steps that are used to solve a problem or perform a specific task. For example, algorithms are used in mathematics to solve mathematical problems, such as equations, factoring numbers, and finding approximate solutions.

On the other hand, artificial intelligence algorithms mimic human intelligence to solve problems and make decisions. These algorithms are based on combining mathematics, statistics, and information theory to develop machine learning and data analysis systems. These algorithms are used in various applications, including robotics, natural language processing, deep learning, data analysis, and optimization.

There are various types of Artificial Intelligence

algorithms, such as supervised and unsupervised learning algorithms, clustering algorithms, neural network algorithms, and genetic algorithms. Each of these algorithms is based on a specific methodology and set of techniques to solve a particular type of problem.

In this sub-chapter, we will analyze the functioning of the main Artificial Intelligence algorithms that have been applied in this field, including Linear Regression, Decision Trees, Neural Networks, and others.

Particle Swarm Optimization (PSO) Algorithm

The PSO algorithm is an optimization method used in artificial intelligence to find the optimal solution to a specific problem. This algorithm is based on the behavior of particle swarms in the solution search space. Particle swarm optimization is a global search technique, meaning it is used to solve problems in which the best possible solutions must be found within a huge search space.

The functioning of the PSO algorithm is based on the simulation of a particle swarm that moves in the solution search space. Each particle represents a possible solution, and its position in the search space represents the characteristics of that solution. The algorithm aims to find the position in the search space where the optimal solution is located.

Each particle is guided by its best previous solution and the best solution found so far by the swarm. The velocity of each particle is adjusted based on this information and is used to move the particle in the search space. The idea is that the particles will move toward the position representing the optimal solution over time.

The PSO algorithm effectively solves problems with many variables and a huge search space. In addition, the PSO is a robust algorithm and can find optimal solutions with moderate computational efficiency. For these reasons, it is a valuable tool in various applications, including optimization problems in engineering and finance.

Genetic Algorithm (GA)
The GA is an optimization search technique based on the theory of evolution and natural selection proposed by Charles Darwin. This algorithm solves complex problems in various fields, including engineering.

The process of operation of the genetic algorithm includes three main stages: initialization, selection, and crossover. An initial population of solutions, also known as chromosomes, is generated in initialization. Each solution represents a possible result for the problem being addressed.

In the second stage, selection, the fitness of each solution in the population is evaluated, and the most fit solutions are selected for use in the next generation. This selection is based on a fitness function, which measures the quality of each solution.

In the third stage, crossover, the selected solutions are combined to generate new individuals, which are also evaluated by the fitness function. This process repeats until an optimal solution is found or a predetermined number of generations is reached.

In addition to selection and crossover, the genetic algorithm also includes a mutation operation, which introduces new solutions to the population. This operation helps prevent the population from getting stuck in a local minimum and increases the diversity of the population.

In summary, the genetic algorithm is a powerful and versatile technique that optimizes complex problems in engineering and other fields. Moreover, with its focus on evolution and natural selection, this algorithm is an excellent choice for solving problems that do not have clear optimal solutions or are difficult to solve using other approaches.

Artificial Neural Network Algorithm (ANN)
The ANN is one of Artificial Intelligence's most essential and widely used. This algorithm is based on the human brain's structure and functioning of neural networks. Artificial neural networks are made up of many nodes or "neurons" connected to each other, and each is designed to process information and perform a specific task.

The functioning of an artificial neural network begins with the input of a signal or data, which is processed by each of the neurons. Next, each neuron in the network performs a mathematical calculation on the input information and sends a signal to the next neuron. This process continues until the signal reaches the network's last neuron, producing an output.

The ANN algorithm is highly flexible and adaptable, meaning it can be trained to perform a wide variety of tasks, from pattern recognition to complex decision-making. In addition, one of the advantages of ANN algorithms is their ability to learn and improve over time. The neural network can adjust its weights and mathematical calculations based on the training data and errors in prediction, leading to continuous improvement in accuracy and effectiveness.

In summary, the ANN algorithm is a powerful and versatile tool in the field of Artificial Intelligence. It has been widely used in various applications, including pattern recognition, decision-making, and optimization.

Algoritmo de Machine Learning (ML)
As experts in Artificial Intelligence, it is essential to understand how one of the most popular algorithms in the industry works: the Machine Learning Algorithm (ML).

The ML Algorithm is based on the idea that a machine can learn independently without being explicitly programmed. Instead of programming a specific solution for a problem, the machine is provided with a dataset. As a result, it can learn independently, adjusting and improving its solutions as it receives more information.

The Machine Learning Algorithm works through three key stages: training, validation, and testing. The machine has a dataset to learn and adjust during the training stage. During the validation stage, the accuracy of the solutions generated by the machine is checked and further adjustment is made if necessary. Finally, the trained model is used to predict new solutions during the testing stage.

The Machine Learning Algorithm uses supervised, unsupervised, and reinforcement learning techniques to improve its accuracy and solutions. In supervised learning, both the data and the correct solutions are provided. In contrast, in unsupervised learning, only the data is provided, and the machine must find patterns and solutions on its own. In reinforcement learning, the machine receives rewards or punishments based on its performance, allowing it to learn and improve continuously.

In conclusion, the Machine Learning Algorithm is a powerful and versatile tool for solving complex problems in a wide range of industries. With its ability to learn independently, it has become one of the most important technologies in Artificial Intelligence and continues to develop quickly.

Finite Element Analysis Algorithm (FEA)
The FEA is a widely used structural engineering

analysis technique for predicting structures' behavior under specific loads. This technique allows engineers to estimate a structure's deformation, stresses, and deflections before building it, which can help identify potential design problems and optimize the structure.

The finite element analysis algorithm creates a digital representation of a structure and divides it into small elements or "cells." Then, these elements are analyzed using a set of mathematical equations that describe their behavior under different loads. Based on these analyses, the algorithm can produce a visual representation of the structure's deformation, stresses, and deflections.

There are many ways to implement the finite element analysis algorithm, from solving differential equations to particle swarm optimization. The choice of technique will depend on the project requirements and the goals to be achieved.

The finite element analysis algorithm is a valuable tool for engineers, as it allows them to predict the behavior of structures in a controlled environment before they are built. This allows potential design problems to be identified and the structure to be adjusted to improve performance. Additionally, the ability to visualize a structure's deformation, stresses, and deflections allows engineers to understand better how a structure works and make informed decisions about its design and construction.

2.3 Challenges in Implementing Artificial Intelligence in Engineering

Implementing AI in engineering presents a series of challenges that must be addressed to ensure effective and successful integration. These challenges include:

2.3.1 Lack of Knowledge and Experience in Artificial Intelligence

AI is a relatively new technology, and many engineers are unfamiliar with its application in their field. Therefore, engineers need to have a solid understanding of the basic concepts of AI and its techniques to effectively apply them in the field of engineering. Additionally, it is necessary to understand machine learning algorithms and their applications in simulation and mathematical calculation.

In many cases, engineers may feel intimidated by the complexity of AI concepts, which can make them reluctant to adopt these technologies. Therefore, providing adequate education on this topic is critical to ensure that engineers have the confidence and skills to implement AI in their projects effectively.

Another challenge in implementing AI in engineering is the lack of adequate tools and resources. Therefore, it is essential to develop specific AI tools for engineers so that they can easily use them in their projects. Additionally, adequate access to the necessary resources and data for training and training AI models is required.

To overcome these challenges, collaboration between AI experts and engineers is necessary to ensure that concepts and tools are developed appropriately and usefully for application in engineering. In addition, investment in research and development in this area is important to ensure that an effective solution is provided to these challenges and that the potential of AI in engineering is maximized.

2.3.2   Incomplete or Poor Quality Data

The lack of data or the presence of incomplete or poor-quality data can be a significant challenge in implementing AI in engineering. This is because AI depends on accurate and high-quality data to perform its work effectively.

AI can only learn and improve over time with accurate and high-quality data. Furthermore, poor data quality can result in the formation of faulty models that can produce incorrect results or even harm decision-making in an engineering project.

Therefore, ensuring that the data used in AI implementation is accurate, complete, and high-quality is essential. This can be achieved through data validation and cleaning, as well as incorporating supervised and unsupervised learning techniques to improve the accuracy of results.

Additionally, it is essential to note that the data used in AI may change over time, so it is necessary to update it regularly to maintain the accuracy and effectiveness of the implementation.

In summary, managing accurate and high-quality data is a key challenge in implementing AI in engineering. It requires a careful and systematic approach to ensure the success of the implementation.

2.3.3   Challenges in Integration with Existing Systems

Integrating AI in engineering requires thorough planning and analysis of existing systems to ensure compatibility and proper architecture. It is also essential to consider the existing systems' ability to handle and process the amount of AI-generated data.

It is also necessary to consider the interoperability between existing systems and AI to ensure an efficient flow of information and seamless interaction. Another challenge in integrating AI in engineering is the need for appropriate hardware infrastructure to host and run AI algorithms. This may require a significant investment in new equipment and hardware, which can be a hurdle for some organizations.

In addition, integrating AI in engineering may require restructuring existing work processes to leverage AI's capacity fully. This can be a challenge regarding time and resources, as it may require a significant reorganization of departments and work teams.

In conclusion, integrating AI in engineering requires careful consideration of existing systems, hardware infrastructure, and work processes. A detailed and strategic approach is essential to ensure successful integration and maximize the unknown findings and solutions that can be obtained through AI in engineering.

2.3.4   Data Protection and Privacy

Data protection and privacy are critical aspects to consider when implementing AI in engineering. AI requires large amounts of data to function effectively, posing significant data protection and privacy challenges. Therefore, engineers must implement solid security measures to ensure the protection of data and user privacy. This may include encrypting sensitive data, implementing rigorous access and authorization controls, and conducting periodic audits to ensure the integrity of security systems.

Moreover, AI often relies on complex algorithms that can have vulnerabilities or be manipulated to produce undesired results. Therefore, ensuring that AI algorithms are secure and not vulnerable to external attacks is essential. Therefore, engineers must continuously evaluate algorithms to identify possible weaknesses and take measures to correct them.

User privacy is also an important aspect to consider. AI may collect and use user data to improve its algorithms, but it is vital to ensure that this data is handled responsibly and that users have control over it. Therefore, engineers must implement clear and strict data protection and privacy policies for users to ensure that their rights are respected, and their personal information is protected.

In conclusion, data protection and privacy are important challenges to consider when implementing AI in engineering. Engineers must be aware of these challenges and take concrete measures to ensure the protection of data and user privacy.

2.3.5 Ethical Implications in Terms of Equity, Diversity, and Non-Discrimination

These ethical challenges must be addressed through appropriate regulation and careful consideration of ethical implications in each AI application. These ethical challenges include, but are not limited to:

Discrimination - ensuring that AI algorithms do not discriminate against certain groups and are fair and equitable in their decision-making.

Bias - ensuring that AI algorithms are not biased towards specific groups and that data used to train AI algorithms is diverse and representative of different perspectives.

Responsibility - determining who is responsible for

the actions of AI algorithms and their impact on society.

Transparency - ensuring that AI algorithms are transparent, and their decision-making processes can be understood and monitored.

Addressing these ethical challenges through appropriate regulation and careful consideration of ethical implications in each AI application is essential. Engineers must be aware of these challenges and take responsibility for ensuring that AI is used in an ethical and equitable manner.

It is important to address the lack of knowledge and experience in AI, resolve the challenges in integrating with existing systems, implement strong security measures to protect data and user privacy and address ethical challenges through proper regulation and careful consideration of ethical implications in every AI application.

UNKNOWN PATTERNS AND SOLUTIONS

I n engineering, it is essential to identify patterns and find efficient solutions to technical challenges. With the advancement of technology, AI has become a valuable tool in the search for unknown patterns and solutions. In this chapter, we will explore in detail how AI can be used to identify patterns and find solutions to complex problems in engineering.

This chapter is crucial in understanding how AI can be used to identify patterns and find unknown solutions in engineering. So, get ready to learn about how AI can revolutionize the way problem-solving is approached in engineering!

3.1   Pattern Descriptors

In engineering, it is crucial to identify and

understand the patterns that develop in systems. This is especially important for improving efficiency and effectiveness in processes. AI plays a pivotal role in discovering unknown patterns and solutions, and in this subtopic, we will focus on pattern descriptors.

Pattern descriptors are an essential tool for identifying and understanding patterns in data. These descriptors provide a compact and meaningful representation of the data, allowing AI algorithms to effectively identify and find unknown patterns and solutions.

There are many types of pattern descriptors, each with strengths and weaknesses. Some of the most common types of pattern descriptors include feature-based descriptors, structure-based descriptors, and topology-based descriptors.

Feature-based descriptors use the intrinsic features of the data to represent it. For example, they may represent the data by using mean, standard deviation, frequency, and other measures. This descriptor type helps identify patterns in numerical data, such as sensor data.

A typical example of feature-based descriptors is the ones used in image analysis. Here, features such as shape, size, color, and texture, among others, can be identified to describe an object in an image. This way, this information can be used to compare and classify different objects in an image, which can be helpful in applications such as object detection or image segmentation.

Another important example is the use of feature-based descriptors in audio signal analysis. Here, features such as frequency, amplitude, and duration, among others, can be identified to describe an audio signal. This way, this information can be used to compare and classify different audio signals, which can be helpful in applications such as sound identification or vowel detection.

It's important to note that selecting the appropriate features to describe an object or system is crucial in the pattern recognition process. Poor feature selection can result in poor pattern recognition and classification, so careful analysis and selection is necessary to ensure the effectiveness of the process.

In summary, feature-based descriptors are an essential tool in the process of identifying and classifying patterns, and their correct implementation can significantly improve the effectiveness and efficiency of analysis and simulation processes in engineering.

Structure-based descriptors are an effective way to represent complex patterns in AI Engineering. This type of descriptor is based on identifying specific structural features of the pattern, including its shape, size, orientation, and position in space.

For example, in medical image analysis, structure-based descriptors can represent tissue patterns, such as cell shape and size or lesion characteristics. In manufacturing, structure-based descriptors can be used to describe patterns in the geometry of parts, such as the shape and size of slots or holes.

Additionally, structure-based descriptors can be combined with other descriptors, such as texture-based descriptors, to improve the representation of complex patterns. For example, structure-based and texture-based descriptors can be combined in defect detection in materials to identify patterns representing defects on the material's surface.

In conclusion, structure-based descriptors are an important tool in AI Engineering for representing complex patterns. This type of descriptor allows for an efficient and precise representation of the patterns, leading to better analysis and problem-solving in a wide range of industrial and scientific applications.

Topology-based descriptors. This subtopic refers to the use of topology in identifying and representing patterns in AI engineering. Topology is a branch of mathematics that studies the shape and structure of mathematical objects, especially when deformed. Topology can be used to identify patterns in data by creating topology-based descriptors.

An example of a topology-based descriptor is the shape descriptor. This descriptor refers to the physical shape of an object and can be used to identify patterns in engineering data. For example, in the automotive industry, the shape of a car part can be used as a descriptor to identify patterns in failure data.

Another example is the algebraic topology

descriptor. This descriptor refers to the mathematical structure of an object and can be used to identify patterns in engineering data. For example, in the aerospace industry, the structure of an aircraft can be used as a descriptor to identify patterns in performance data.

Furthermore, topology-based descriptors can also be used to identify patterns in process engineering data. For example, in the petrochemical industry, the structure of a refining process can be used as a descriptor to identify patterns in production data.

In summary, topology-based descriptors are a valuable tool in the identification and representation of patterns in AI engineering and can be used in a variety of applications, including the identification of patterns in product, process, and performance engineering data.

It is important to note that the selection of appropriate pattern descriptors and their optimization are crucial for the success of AI algorithms in engineering. An inadequate selection of pattern descriptors can result in poor data representation and, therefore, in inaccurate or even incorrect results. Consequently, it is essential to understand how pattern descriptors work and how to optimize them to obtain the best results.

3.2 Pattern Detection with Artificial Intelligence (Algorithms)

In this subtopic, we will explore how AI can be used to detect patterns in large amounts of data. Pattern

Detection is critical in engineering as it allows engineers to identify trends, relationships, and cause-and-effect relationships that might otherwise go unnoticed.

AI-based pattern detection uses machine learning techniques to analyze large amounts of data and detect patterns that can be used to improve the efficiency and effectiveness of processes and systems. Some of the techniques used in AI-based pattern detection include deep learning, neural networks, and clustering analysis.

3.2.1   Deep Learning
Allow AI algorithms to learn to detect patterns in data by using neural networks. This technique uses interconnected node layers to analyze data and see patterns.

Deep learning is one of the most advanced pattern detection techniques in AI. This technique uses artificial neural networks to model and learn complex relationships between data. The idea behind deep learning is that as the neural network is fed more data, it becomes increasingly accurate in pattern identification.

This type of learning differs from other AI techniques because the neural networks learn independently, without the engineer's specific programming. The neural network uses multiple layers of nodes and decision trees to identify patterns and relationships in data, making it highly effective in processing large amounts of data and in identifying complex patterns.

Some names of algorithms used in this technique are Artificial Neural Networks, Convolutional Neural Networks (CNN), Recurrent Neural Networks (RNN), Deep Belief Networks (DBN), Autoencoder, Generative Adversarial Networks (GAN), Long Short-Term Memory (LSTM), Self-Organizing Maps (SOM), Restricted Boltzmann Machine (RBM), Convolutional Deep Belief Network (CDBN).

An example of a deep learning application in engineering is detecting faults in mechanical components. The neural network can be trained with data from past failures, allowing the network to identify patterns in component data that correlate with future failure. In this way, the neural network can help predict failures before they occur, preventing serious and costly damage.

Another example of deep learning application in engineering is optimizing production processes. The neural network can be trained with past production data, allowing the network to identify production data patterns that correlate with higher efficiency. In this way, the neural network can help improve production processes to increase efficiency and reduce costs.

3.2.2   Neural Networks

They are a fundamental tool in pattern detection with AI. The functioning of the human brain inspires these structures and allows for complex automatic learning tasks. Neural networks consist of nodes, which represent brain cells, and connections, which represent brain synapses. Each node processes information and sends a signal to other nodes, which in turn process the information and send new alerts. Through training, neural networks can learn to identify patterns and solutions in complex data.

Some names of algorithms used in this technique are Backpropagation Algorithm, Multilayer Perceptron Algorithm, Forward Propagation Algorithm, Convolutional Neural Network (CNN) Algorithm, Recurrent Neural Network (RNN) Algorithm, Long Short-Term Memory (LSTM) Algorithm, Deep Belief Network (DBN) Algorithm, Generative Adversarial Network (GAN) Algorithm, Neural Network Autoencoder Algorithm, Boltzmann Machine Algorithm.

Some examples of neural network applications in pattern detection include:

1. Prediction of industrial component failure: neural networks can be used to analyze patterns in historical data of industrial components and accurately predict when a piece will fail.

2. Medical diagnosis: Neural networks can be used to analyze medical images and make accurate diagnoses.

3. Fraud detection: neural networks can be used to detect fraud in financial transactions by analyzing patterns in

transaction data.

4. Image and speech recognition: neural networks can be used for image and speech recognition by analyzing image and speech data patterns.

These are just a few examples of how pattern recognition with AI can be applied in various engineering fields. Using neural networks allows for the identification of complex patterns and the generation of efficient and accurate solutions.

3.2.3 Clustering Analysis

It is an AI technique that allows for grouping similar objects into the same group or cluster. This technique is used to identify patterns and trends in large amounts of data and improve understanding.

In clustering analysis, objects are evaluated based on their characteristics and assigned to a specific cluster based on their similarity. The number of clusters that need to be formed is an important decision that is taken before starting the analysis and can be determined through different methods, such as the variance analysis or the elbow method.

Clustering can be performed through different algorithms, such as K-means, Hierarchical Clustering, DBSCAN, Expectation-Maximization (EM), Fuzzy C-means (FCM), Self-Organizing Maps (SOM), Spectral Clustering, Gaussian Mixture Model (GMM), Agglomerative Clustering, Birch Clustering Algorithm. Each algorithm is suitable for different data types and objectives, so choosing the appropriate algorithm for the problem to be solved is vital.

Some examples of how clustering analysis can be used include:

1. Market segmentation: Grouping customers based on their characteristics and preferences to better understand their needs and offer personalized products or services.
2. Trend analysis: Identifying patterns in large amounts of data, such as consumer preferences, to make informed decisions in marketing and advertising.
3. Medical diagnosis: Grouping patients with similar symptoms and diseases for a better understanding of the pathology and a more effective treatment.
4. Security analysis: Grouping similar events in a network to identify patterns and trends in cybersecurity.

These are just a few examples of how pattern recognition with AI can help engineering improve its processes and discover unknown solutions. Using clustering techniques, engineering can transform data into valuable information and use it to solve problems more efficiently and accurately.

Pattern recognition with AI has several applications in engineering, including improving energy efficiency, detecting anomalies, and optimizing processes. By using advanced AI techniques to detect patterns in data, engineers can make informed decisions and improve the efficiency and effectiveness of systems and processes.

In conclusion, pattern recognition with AI is a critical technology in engineering that allows engineers to identify patterns, trends, and causal relationships in data. By using advanced AI techniques, engineers can make informed decisions and improve efficiency and effectiveness.

3.3 Unknown Solutions and Artificial Intelligence

In engineering, AI solves complex and challenging problems that have traditionally been difficult to solve using traditional mathematical methods. Some examples might be:

3.3.1 Artificial Intelligence in Urban Planning and Infrastructure Construction

Artificial Intelligence can analyze large amounts of data and model future scenarios to help engineers make informed decisions about the location and design of infrastructure. In this field, AI can optimize the location of new infrastructure and improve city traffic flow. For example, AI can analyze traffic and demographic patterns to determine the optimal location for new infrastructure, such as hospitals, schools, and shopping centers. Additionally, AI can be used to optimize the design of transportation systems such as roads and metro lines to improve traffic flow and reduce travel times.

AI can also use optimization techniques to evaluate different construction scenarios and determine the best option. For example, AI can simulate different design and construction options for a road and determine the option that offers the best solution in terms of cost, efficiency, and safety. Another example of the application of AI in urban planning and infrastructure construction is using drones and robots to inspect and monitor construction. AI can be used to analyze images and data captured by drones and robots to evaluate construction progress and detect any issues in real time.

Example: "Traffic pattern analysis with the help of AI"

AI can be used to analyze traffic patterns in a specific area and produce a visual representation of the distribution and movement of vehicles. Additionally, AI can be used to detect and prevent traffic congestion. AI can analyze information collected by traffic sensors and provide a real-time representation of traffic congestion in a specific area. This allows traffic planning authorities to take preventive measures to reduce traffic congestion. AI can also be used to improve traffic efficiency and optimize vehicle routes. AI can analyze traffic and demographic patterns to optimize vehicle routes and improve traffic efficiency in a specific area. This allows drivers to reach their destinations more quickly and reduces traffic congestion on the road. In summary, AI is a valuable tool for traffic pattern analysis because it can process and analyze large amounts of data in real-time. As a result, AI can be used to detect and prevent traffic congestion, improve traffic efficiency, and optimize vehicle routes, which can help improve driver safety and reduce traffic congestion on roads.

3.3.2   Evaluation of Landslide and Hillside Slip Risks

AI can analyze large amounts of geological and climatic data, including soil composition, topography, precipitation, erosion, and vegetation information. These data are combined with mathematical models and machine learning algorithms to predict the probability of a landslide or hillside slip in a specific area. AI can also use monitoring techniques to detect the early signs of a landslide or hillside slip and provide warnings to those affected. By using AI to evaluate landslide and hillside slip risks, authorities can make more informed decisions about the safety and stability of slopes and take preventive measures to reduce the risk of landslides and hillside slips. For example, if a high-risk landslide zone is identified, engineers can take steps to strengthen the soil structure and prevent a disaster.

Example: "Analysis of climatic and geological patterns to more accurately predict the probability of a landslide."
This is achieved using machine learning algorithms and advanced data analysis techniques. To carry

out this analysis, relevant climatic and geological data, such as precipitation, temperature, humidity, and the geological characteristics of the terrain, are collected and integrated. These data are combined with information on past landslides and used to train an AI model. Once trained, the AI model can use climatic and geological patterns to predict the probability of a landslide in a specific area. To make precise predictions, these models can consider multiple factors, such as topography, soil permeability, and erosion. In addition, the AI model can also consider dynamic factors such as changes in the climate and vegetation and adjust its prediction accordingly. These results can be used to take preventive measures and reduce disaster risk in the area. For example, strategies can be implemented to improve soil stability, such as constructing barriers or relocating buildings to low-risk landslide areas. AI can also be used to continuously monitor the situation and detect any signs of an imminent landslide event. This allows authorities to take rapid and effective measures to prevent damage to property and people.

3.3.3 Construction of Bridges and Dams to Improve Project Efficiency and Safety

AI can be used to analyze real-time traffic patterns and static and dynamic loads, allowing engineers to make informed decisions about the design and construction of these critical projects. One of the ways AI can improve the construction of bridges and dams is through load distribution optimization. AI can use algorithms to analyze dynamic and static loads and optimize the load distribution in the structure to minimize the risk of failures and ensure safety. In addition, AI can analyze the structure's geometry to ensure it complies with safety standards and to identify any weaknesses that must be corrected. Another way AI can improve the construction of bridges and dams is through real-time construction monitoring. AI can use sensors and cameras to monitor the building in real-time and detect any issues or anomalies. This allows engineers to respond to problems quickly and efficiently before affecting the structure's safety or integrity.

Example: "Load distribution optimization with AI"
Optimizing load distribution in a structure is a key factor in ensuring its safety and efficiency. AI can help optimize this load distribution efficiently and accurately. The process of load distribution optimization starts with creating a virtual model of the structure. This virtual model is fed with relevant data such as the structure's geometry, material properties, applied loads, and design criteria. Then, an AI algorithm is used to analyze the virtual model and optimize the load distribution. This algorithm can be a heuristics-based optimization algorithm or a machine learning algorithm, depending on the complexity of the problem and the optimization objectives. The AI algorithm considers

various factors to optimize the load distribution, such as energy efficiency, safety, strength, and the structure's durability. It also considers design constraints and limitations. As a result, load distribution optimization using AI can result in a safer and more efficient structure as it reduces the structure's weight and total costs, increasing its lifespan and improving resistance to loads and natural disasters.

In summary, AI is a valuable tool for engineering that allows engineers to discover unknown solutions and optimize processes in infrastructure construction, risk assessment, and bridge and dam construction. This third subtopic delves into how AI can help civil engineering overcome complex challenges and improve the safety and efficiency of projects.

AUTOMATED LEARNING SYSTEMS AND THEIR APPLICATION IN ENGINEERING

These systems allow AI to learn from data and improve its performance over time.

Automatic learning systems are divided into two main categories: supervised and unsupervised. In supervised learning, AI is fed with previously labeled data which is used to train the system. In unsupervised learning, AI works with unlabeled data and searches for patterns or relationships in these data to be able to learn.

The application of automatic learning systems in

engineering is extensive. For example, they can be used to analyze factory production data patterns, predict equipment downtime, and optimize production processes.

Automatic learning systems use AI algorithms and techniques to learn from data and make predictions and analyses. In civil engineering, these systems can be used to analyze patterns in construction, design, and urban planning data and predict structures' problems and failures. These systems can also be used to simulate and analyze the behavior of structures under different weather and load conditions, allowing civil engineers to make informed decisions about the design and construction of facilities.

Additionally, automatic learning systems can be used to optimize construction processes and improve efficiency in terms of time and cost. These systems can analyze large amounts of data and perform complex calculations, allowing engineers to make data-based decisions instead of relying on assumptions and suggestions.

In summary, the application of automatic learning systems in civil engineering can be very useful for improving current methods and processes and optimizing structures' construction and design. This chapter of the book "AI in Engineering: Discovering Unknown Patterns and Solutions" explores in depth how automatic learning systems can be used in civil engineering and how they can help engineers improve the efficiency and quality of their projects.

4.1   Linear Regression and Its Application in Engineering

The linear regression method is one of AI's most widely used machine learning systems. This method seeks to establish a linear relationship between two or more variables, aiming to predict or estimate a dependent variable based on one or more independent variables.

In engineering, this method is applied in various areas, such as predicting energy demand, estimating the lifespan of components and materials, and estimating the number of materials needed for a construction project.

For example, in civil engineering, the linear regression method can be used to predict the number of materials needed to build a building based on the construction surface area and the number of floors, among other factors. It can also be used to estimate energy demand in a building, based on the number of inhabitants and the outdoor temperature.

Another example is mechanical engineering, where linear regression can be used to model the relationship between input variables and material properties. This can help improve manufacturing processes' efficiency and the quality of final products.

In addition to the above applications, it can

also be used for process optimization in industry. For example, in chemical engineering, machine learning systems can be used to model the performance of a process and predict its future performance. It can also be used to optimize product quality and reduce production costs.

It is important to note that the linear regression method is the only tool within the wide range of machine learning systems that can be applied in engineering. However, its simplicity and ability to handle large data make it especially useful for many engineering problems.

4.2   Decision Trees and Their Application in Engineering

The decision tree method is a widely used machine learning algorithm in the industry. This method is based on creating a decision model that can classify or predict values based on a data set. Decision trees are a handy tool in engineering, as they allow engineers to make decisions based on patterns and trends in the data.

Decision trees can be used in civil engineering for material selection in building and structure construction. The material selection process in construction is a critical aspect, as materials directly affect safety, durability, and the project's overall cost. Therefore, it is crucial to make informed decisions about material selection.

The decision tree method can be used to identify important criteria for material selection, such as strength, durability, availability, and cost. Based on these criteria, branches can be created in the decision tree representing different material options. In addition, each branch can have its own sub-criteria and branches, allowing for a more detailed and exhaustive evaluation of the available options.

Once all the material options have been evaluated, the decision tree method can help identify the most suitable material for the project based on the specific features and requirements. This method can be more efficient and accurate than traditional methods, providing a clear and concise way to make informed decisions.

Decision trees can also be used in urban planning to predict traffic congestion and demand for public services, such as hospitals, schools, and emergency services. These models can help urban planners to identify areas with higher congestion and develop solutions to improve residents' quality of life.

Building a decision tree starts with selecting a set of relevant variables, such as population density, the number of homes, the location of public points of interest, and the terrain topography. Then, a machine learning algorithm is used to identify patterns and relationships between these variables and build a decision tree that represents these patterns.

Engineers can use this information to predict traffic congestion in a city, identifying areas with higher demand for public services, such as hospitals, schools, and emergency services. This information is essential for urban planning, as it allows engineers to design more efficient and optimized infrastructure that responds adequately to the population's needs.

In addition, decision trees can be updated and improved over time using updated information and new relevant variables. This allows engineers always to be aware of the latest trends and changes in traffic congestion and demand for public services and respond accordingly in urban planning.

In conclusion, the decision tree method in machine learning is a powerful tool in engineering, allowing engineers to make decisions based on patterns and trends in data, improving the efficiency and quality of engineering solutions to problems.

4.3   Neural Networks and Their Application in Engineering

The decision tree method is a widely used machine learning algorithm in the industry. This method is based on creating a decision model that can classify or predict values based on a data set. Decision trees are a handy tool in engineering, as they allow engineers to make decisions based on patterns and trends in the data.

Decision trees can be used in civil engineering for material selection in building and structure construction. The material selection process in construction is a critical aspect, as materials directly affect safety, durability, and the project's overall cost. Therefore, it is crucial to make informed decisions about material selection.

The decision tree method can be used to identify important criteria for material selection, such as strength, durability, availability, and cost. Based on these criteria, branches can be created in the decision tree representing different material options. In addition, each branch can have its own sub-criteria and branches, allowing for a more detailed and exhaustive evaluation of the available options.

Once all the material options have been evaluated, the decision tree method can help identify the most suitable material for the project based on the specific features and requirements. This method can be more efficient and accurate than traditional methods, providing a clear and concise way to make informed decisions.

Decision trees can also be used in urban planning to predict traffic congestion and demand for public services, such as hospitals, schools, and emergency services. These models can help urban planners to identify areas with higher congestion and develop solutions to improve residents' quality of life.

Building a decision tree starts with selecting a set of relevant variables, such as population density, the number of homes, the location of public points of interest, and the terrain topography. Then, a machine learning algorithm is used to identify patterns and relationships between these variables and build a decision tree that represents these patterns.

Engineers can use this information to predict traffic congestion in a city, identifying areas with higher demand for public services, such as hospitals, schools, and emergency services. This information is essential for urban planning, as it allows engineers to design more efficient and optimized infrastructure that responds adequately to the population's needs.

In addition, decision trees can be updated and improved over time using updated information and new relevant variables. This allows engineers always to be aware of the latest trends and changes in traffic congestion and demand for public services and respond accordingly in urban planning.

In conclusion, the decision tree method in machine learning is a powerful tool in engineering, allowing engineers to make decisions based on patterns and trends in data, improving the efficiency and quality of engineering solutions to problems.

SIMULATION AND MATHEMATICAL CALCULATION MODELS WITH ARTIFICIAL INTELLIGENCE

I n engineering, simulation and mathematical calculation models are vital tools for predicting and understanding the behavior of systems and processes. However, these traditional models usually rely on a combination of mathematical equations and empirical values. As a result, they are limited in their ability to process large amounts of data and solve complex problems. AI can be a valuable complement to these models by using machine learning techniques to improve the accuracy and efficiency of the calculations.

One way AI can be applied to simulation and mathematical calculation models is through optimization techniques. For example, genetic algorithms and ant colony algorithms can be used to optimize the variables and parameters in a model, resulting in more accurate and efficient solutions.

Another way that AI can be used in simulation and mathematical calculation models is by creating predictive models. For example, machine learning algorithms such as neural networks can be trained with large amounts of historical data to predict the future behavior of a system. These predictive models can be used to improve planning and decision-making in engineering.

In the engineering world, accuracy and detail are crucial to ensuring safety and efficiency in the design and construction of structures. To achieve this, AI techniques can be used to develop more accurate and detailed mathematical models for simulation and analysis.

An example of this is using artificial neural networks to simulate the response of a structure to various stimuli and external factors, such as wind, earthquake, or load. These systems are based on modeling the structure as a network of nodes connected, each representing a specific part of the structure. The neural network is trained using real data on the structure's response to different stimuli, and from this information, it can predict the structure's response to new stimuli.

In simulating a structure's response to wind, for example, the neural network can be trained using data on wind speed and direction, as well as the structure's response to different wind levels. From this data, the neural network can predict how the structure will respond to new wind levels, allowing engineers to predict and prevent potential failures.

Similarly, in simulating a structure's response to an earthquake, the neural network can be trained using data on the intensity and direction of the earthquake, as well as the structure's response to different earthquake magnitudes. From this data, the neural network can predict how the structure will respond to new earthquakes, allowing engineers to evaluate the seismic resistance of a structure and take measures to improve it.

5.1 Dynamic Systems Simulation Models

Dynamic systems simulation models are a key tool in engineering that allows engineers to predict and understand the behavior of a system over time. These models use mathematics and physics to describe the relationships between variables and the system's dynamics. AI can be used to improve these dynamic systems simulation models in several ways.

On the one hand, AI can identify patterns and trends in the data collected about the simulated dynamic system. This data analysis can help engineers better understand the interactions between variables and optimize the simulation models. For example, if a power transmission system is being simulated, AI can help identify patterns in energy demand that can be used to improve the system's efficiency and stability.

AI can also be used to improve the accuracy and efficiency of dynamic systems simulation models. For example, AI algorithms can optimize the simulation models' parameters, allowing for better prediction of the system's behavior over time. This is particularly useful when dealing with dynamic systems with many variables and complex interactions, such as an urban transportation system.

Finally, AI can be used to develop new dynamic systems simulation models that allow for a better understanding of the phenomena being simulated. For example, simulation models based on artificial neural networks can be developed to simulate a system's response to different stimuli and external factors, such as wind, earthquakes, or load. These models can be used to predict the system's behavior in extreme situations, which is crucial for infrastructure safety.

In summary, the use of dynamic systems simulation models combined with AI techniques represents a significant advancement in the field of engineering. It allows for a better understanding and optimization of the systems and processes studied. This combination of tools enables engineers to solve more complex and challenging problems more efficiently and accurately.

5.2   Mathematical Calculation Models with Artificial Intelligence

Mathematical calculation models with AI are valuable in structural engineering, as they allow for more efficient and accurate simulation and analysis of complex systems than traditional methods. AI can perform complex mathematical calculations, identify patterns, and make predictions based on previous data. This allows structural engineers to develop more accurate and detailed models to simulate a structure's response to different stimuli and external factors.

For example, in structural engineering, the similarity of vibrations and deformations of a structure is a critical aspect of its safety and durability. In addition, most tall buildings, due to their height and weight, are subject to dynamic loads such as wind and earthquakes. Therefore, it is essential to accurately simulate and analyze a structure's response to these external stimuli to ensure its stability and safety.

Traditional structural analysis methods, such as finite element methods, require many calculations and time to simulate a structure's response. However, applying AI in structural engineering allows these responses to be simulated more accurately and efficiently.

An example of this is the use of linear regression

algorithms in simulating a structure's response to different stimuli. Linear regression is a machine learning technique used to model the relationship between two or more variables. Using these models, a structure's responses to different stimuli, such as wind and earthquakes, can be simulated with greater accuracy and speed than traditional methods.

In addition, AI techniques, such as artificial neural networks, allow for modeling a structure's response to different stimuli and external factors. These models use an artificial neural network to simulate a structure's response to external stimuli, allowing for a more accurate and detailed simulation of a structure's response to dynamic loads.

In conclusion, using AI in structural engineering allows for the development of more accurate and detailed simulation models to simulate a structure's response to different stimuli and external factors. These models, whether linear regression algorithms or artificial neural networks, allow for better evaluation of the safety and stability of tall buildings, resulting in better planning and construction of high-rise structures.

5.3   Comparison of Traditional Models with AI-Based Models

In the field of structural engineering, simulation models and mathematical calculations are crucial tools for predicting the behavior of structures under different stimuli and external factors. In the past, these models were based on traditional methods such as linear and nonlinear static analysis, dynamic analysis, and finite element analysis. However, with the advent of AI, new AI-based methods have been developed that improve the accuracy and detail of these models.

Traditional models are based on the mathematical resolution of equations that describe the behavior of the structure. These models are helpful, but they have some limitations, such as mathematical complexity, uncertainty in input data, and a lack of detailed information about the behavior of the structure. On the other hand, AI-based models use machine learning techniques to identify patterns in the data and generate more accurate and detailed mathematical models.

The process for identifying patterns in the data and generating AI-based models is outlined below. In the first phase, the necessary information is collected, which can include sensor measurements, and data logs, among others. The next phase consists of data pre-processing, which includes cleaning and normalizing the data to improve its quality.

The next phase is the training of the neural network. A portion of the data is used to teach the model to identify patterns and relationships in the data. This is done by adjusting the weights and trends of the neurons in the network until the model can make accurate predictions.

Finally, in the last phase, the model is evaluated using a data set that was not used during training.

This process is essential to determine the model's accuracy and ensure that it generalizes well to new data.

Identifying patterns and generating AI-based models requires a combination of mathematics, engineering, and information technology skills. In addition, a rigorous and systematic approach is necessary to ensure that the models are accurate and reliable.

In conclusion, AI-based models improve traditional models in terms of accuracy and detail. Although there are still limitations in their application, their potential to improve the prediction of the behavior of structures is evident, and their use is spreading in the field of structural engineering.

DEVELOPMENT OF ARTIFICIAL INTELLIGENCE APPLICATIONS IN ENGINEERING

Artificial Intelligence (AI) is a field of study revolutionizing how complex problems are approached. This technology has been integrated into various industries, from medicine to logistics and engineering.

Engineering is a discipline that requires a deep knowledge of materials, physics, and mathematics to design and build structures and systems that meet technical and safety requirements. AI can be used to simplify the design process and optimize real-time decision-making.

For example, AI models can be developed to

simulate the response of a structure to different stimuli, such as wind, earthquake, or load. These models are based on machine learning algorithms and the identification of patterns in sensor measurement data. In this way, a detailed and accurate representation of the structure's response to different stimuli can be obtained, allowing informed decisions about optimizing its design and construction.

Another aspect where AI can be highly useful is developing simulation models for dynamic systems. These models allow simulation of the response of a system to different stimuli and external factors such as pressure, temperature, or deformation. These models are essential to ensure the safety and proper functioning of systems and structures in real-time.

In addition, AI can also be used to develop more accurate and detailed mathematical calculation models. These models allow for complex calculations to be performed more quickly and efficiently, meaning that more informed and faster decisions can be made. This is especially useful in emergencies where quick and precise decisions are needed to ensure the population's and infrastructure's safety.

6.1 Examples of Artificial Intelligence Applications in Engineering

Some examples of applications in Civil Engineering:

1. Construction process optimization: An AI model can help optimize construction

processes, for example, through resource allocation and the planning of construction material production.

2. Earthquake scenario simulation: An AI application can simulate earthquake scenarios and predict the response of a structure to an earthquake. This can help improve safety in high seismic hazard areas.

3. Construction material identification: An AI application can identify the construction materials used in a structure from an image or 3D scan. This can be useful for more efficient inspection and monitoring of structures.

4. Energy efficiency optimization in buildings: An AI model can analyze patterns in energy use in a building and optimize energy use to improve energy efficiency.

5. Structural life prediction: An AI application can predict the life of structures, allowing engineers to properly plan maintenance and renovation of these.

6. Structural life analysis: AI can be used to simulate the behavior of structures and estimate the remaining life of these, which can help prioritize maintenance and renovation.

7. Failure detection in structures: sensors and AI algorithms can be used to identify patterns in the vibration and deformation of a structure, allowing for failures to be detected before they are visually evident.

8. Load distribution optimization in structures:

AI algorithms can be used to simulate the response of a structure to different load scenarios and find the best load distribution.

9. Safety analysis in structures: AI algorithms can be used to analyze the behavior of structures in case of earthquakes and strong winds, allowing critical points to be identified and safety to be improved.

10. Material selection: AI can assist in selecting materials for building structures, and evaluating factors such as strength, durability, and cost.

11. Three-dimensional modeling of structures: AI algorithms can be used to create three-dimensional models of structures, allowing for better visualization of their behavior under different loads and external factors.

12. Evaluation of structure designs: AI can evaluate different structure designs and determine the most efficient and safe.

13. Prediction of deformation and vibration in structures: AI algorithms can be used to predict the deformation and vibration of a structure, allowing problems to be identified and corrected before they become evident.

14. Real-time monitoring of structures: sensors and AI can be used to monitor the behavior of a structure in real time and detect any issues or deviations in performance.

15. Prediction of lifespan and fatigue in structural elements such as beams, columns, and bridges

using AI techniques such as deep learning and linear regression.

16. Real-time structural safety analysis through constant monitoring of the structure with sensors and the application of AI models to detect any deviations in the shape or behavior of the structure.

17. Automatic identification of structural defects and problems using computer vision and deep learning techniques to improve efficiency and accuracy in structure inspections.

Some examples of application in Road Engineering:

1. Traffic control systems with AI: AI monitors and controls traffic flow in real-time, maximizing efficiency and reducing congestion.

2. Detection and prevention of traffic accidents: AI-equipped vehicles can detect and prevent traffic accidents by analyzing data and identifying patterns.

3. Real-time weather condition analysis: AI systems can collect and analyze weather data in real-time to determine road safety and improve route planning.

4. Prediction of traffic congestion: AI algorithms can predict traffic congestion and recommend alternative routes to drivers.

5. Detection of potential accidents: AI-equipped sensors can detect and alert

drivers to potential accidents on the road.

6. Predictive maintenance: AI systems can analyze and predict the need for maintenance on roads and bridges, allowing for efficient and proactive planning.

7. Improving signal efficiency: AI systems can optimize real-time signaling, improving traffic safety and efficiency.

8. Route and risk analysis: AI systems can analyze routes and identify risks to improve road planning and safety.

9. Prediction of construction project duration: AI algorithms can predict the duration of construction projects and improve planning and budgeting.

10. Construction quality monitoring: AI systems can monitor and evaluate construction quality, ensuring the safety and durability of road infrastructure.

Examples of applications in Mechanical Engineering:

1. Prediction of mechanical equipment failures: using AI algorithms, patterns in performance data and failures can be analyzed to predict when a piece of equipment will fail.

2. Production process optimization: AI algorithms can be used to optimize production processes, improving efficiency,

and reducing costs.

3. Automated quality control: Artificial neural networks can be trained to detect product defects, reducing inspection time and costs.

4. Vibration and deformation analysis: AI algorithms can be used to simulate vibrations and deformations of mechanical components, allowing for a better understanding of their behavior.

5. Load prediction: Artificial neural networks can be trained to predict load on mechanical systems, allowing for a better understanding of their behavior.

6. Automated diagnosis: AI algorithms can be used for automated diagnostics, allowing for early detection of problems.

7. Manufacturing process control: AI algorithms can be used to control manufacturing processes, improving efficiency and quality.

8. Energy efficiency improvement: AI algorithms can be used to improve the energy efficiency of mechanical systems, reducing costs.

9. Resource management optimization: AI algorithms can be used to optimize resource management, improving efficiency, and reducing costs.

10. Control of complex mechanical systems: AI algorithms can be used to control complex mechanical systems, allowing for

a better understanding and control of their behavior.

Examples of applications in Hydraulic Engineering:
1. Fire prediction using machine learning algorithms.
2. Real-time water quality monitoring in rivers and aquifers with sensors connected to neural networks.
3. Flood and dam overflow prediction with AI-based models.
4. Analysis of water consumption patterns in cities to improve efficiency in distribution.
5. Development of a control system for regulating water quantity in an intelligent irrigation system.
6. Monitoring river erosion and sedimentation in dams using drones and AI algorithms.
7. Development of a tracking system for water distribution in cities using AI algorithms.
8. Improving the efficiency of water treatment plants using AI algorithms.
9. Monitoring and predicting soil moisture levels in agriculture using sensors and AI algorithms.
10. Development of a system for monitoring and controlling the water levels in reservoirs with AI algorithms.

Some examples of application in Process Engineering for the oil industry:

1. Predicting equipment and system failures in the oil industry using AI.
2. Analysis of oil and gas flows in transportation pipelines using machine learning algorithms.
3. Real-time monitoring of the quality of oil and gas during transportation to prevent contamination and protect the environment.
4. Automated identification and selection of the most efficient oil and gas extraction processes.
5. Predicting the lifespan of equipment and components in the oil industry to optimize preventive maintenance.
6. Optimization of oil and gas production using AI algorithms to control process variables.
7. Identification of oil and gas transportation pipeline leak sources using AI.
8. Analysis of sensor data in the oil industry to detect patterns and trends that may affect process efficiency.
9. Prediction of the behavior of oil facilities in adverse situations such as storms or earthquakes to ensure people's safety and the environment's protection.
10. Optimization of resource utilization in the oil industry using AI to improve efficiency and reduce costs.

Some examples of applications in Geotechnics:

1. Prediction of landslides using deep learning algorithms.
2. Evaluation of slope stability using data analysis and mining techniques.
3. Real-time monitoring of terrace deformation using intelligent sensors and predictive analysis.
4. Optimization of geological resource extraction using AI techniques.
5. Identification of patterns in soil behavior and its relationship with climatic and geological events.
6. Ground consolidation and material deformation forecast models.
7. Evaluation of soil strength and its relationship with structures built on it.
8. Prediction of soil responses to dynamic loads such as earthquakes or explosions.
9. Analysis of the quality of fill materials for dam and dike construction using AI techniques.
10. Monitoring of the integrity of underground structures and surface structures in the construction of tunnels and underground roads.

Some examples of application in Sanitary Engineering:

1. Predictive demand for potable water: AI is used to predict the demand for potable

water in a given area to ensure an adequate and efficient supply.

2. Water quality monitoring: AI monitors water quality in a distribution network, identifying possible leaks or contamination problems.

3. Optimization of wastewater treatment: AI is used to optimize wastewater treatment processes, increasing efficiency, and reducing costs.

4. Maintenance needs prediction: AI predicts the maintenance needs of water and sewage systems, reducing downtime and unnecessary costs.

5. Analysis of water distribution network performance: AI is used to analyze the performance of water distribution networks, improving efficiency, and reducing losses.

6. Optimization of irrigation systems: AI is used to optimize irrigation systems, improve water use efficiency, and reduce costs.

7. Predicting water shortages: AI is used to predict water shortages in a given area to plan for and prevent water scarcity.

8. Analysis of water consumption patterns: AI is used to analyze water consumption patterns in a given area to identify and address inefficiencies in water usage.

9. Detection of leaks in water distribution

systems: AI is used to detect leaks in water distribution systems, reducing losses and improving system efficiency.

10. Planning and design of water and sewage systems: AI is used to plan and design water and sewage systems, considering various factors such as population growth, urbanization, and environmental impact.

Some examples of applications in Port Engineering:

1. Water quality prediction in ports: AI algorithms can be used to analyze and predict the water quality in ports. This allows engineers to take preventive measures to minimize risks and preserve environmental health.

2. Port activity monitoring: Sensors and cameras can gather real-time information about the port activity, such as boat traffic and cargo loading. AI can be used to analyze this information and predict future traffic to improve efficiency and safety.

3. Space utilization optimization: AI algorithms can be used to optimize space utilization in ports, identifying the best places to store boats and cargo and reducing wait time and storage costs.

4. Construction planning and monitoring: AI algorithms can help engineers plan and monitor port construction, identifying obstacles and optimizing construction

efficiency.

5. Weather and tide prediction: AI algorithms can be used to predict weather and tides in ports, allowing engineers to take preventive measures to minimize risks and preserve safety.

6. Infrastructure monitoring: Sensors can gather information about port infrastructure, such as the condition of piers and cranes. AI can be used to analyze this information and predict possible failures to prevent problems and improve efficiency.

7. Maritime traffic planning and monitoring: AI algorithms can help engineers plan and monitor maritime traffic in ports, optimizing efficiency and safety.

8. Environmental impact prediction: AI can be used to predict the environmental impact of port activity, identify potential pollution sources, and take preventive measures to minimize risks.

9. Resource management optimization: AI algorithms can be used to optimize resource management.

6.2 Process of developing AI applications in engineering

The development of AI applications in engineering is a process that requires a combination of skills in computer science, mathematics, engineering, and industry knowledge in the field where the application is desired. AI is a technology that has the potential to transform the way structures, processes, and infrastructures in engineering are designed, built, and operated.

The process of developing AI applications in engineering can be summarized in the following steps:

1. Define the problem: Identification of the problem: The first stage is to identify a problem that can be solved with the help of AI. Engineering can be a vast field; any specific issue will require deep research to determine if AI is a viable solution.

2. Select the data: Once the problem is defined, collecting and selecting relevant data for application development is necessary. This includes collecting historical data and real-time sensors. This data can be collected from external sources or generated internally.

3. Data preprocessing: The collected data must be preprocessed to remove irrelevant data and improve data quality. This includes cleaning, integrating, and transforming the data.

4. Modeling: In this step, a mathematical or AI model is developed that is capable of solving the problem identified in the first step. This may include selecting an existing

model or creating a custom model.

5. Validation and testing: Once the model is developed, it must be validated to ensure accuracy and reliability. Tests must be conducted using historical data and real-time sensors to evaluate its performance.

6. Implementation and monitoring: Finally, the AI model is integrated into the relevant system and used to solve the original problem. This integration may require modifying the existing system and continuous evaluation to ensure its effectiveness and efficiency.

Developing AI applications in engineering is continuous and requires a rigorous approach and a dedicated team to ensure its success. Nevertheless, AI has the potential to revolutionize engineering and solve complex problems that were once considered impossible. First, however, it is crucial to consider the ethical and technical challenges that arise with the implementation of AI in engineering and be prepared to address them effectively.

6.3 Tools and technologies required for the development of AI applications in engineering.

The development of AI applications in engineering requires a combination of advanced tools and technologies that allow the creation of intelligent systems capable of solving complex problems and optimizing processes. Below, we will describe some of this field's most relevant tools and technologies.

Machine learning: algorithms are the foundation of AI and allow systems to learn and evolve based on the information and data they receive. There are various types of algorithms, such as supervised, unsupervised, and reinforced learning, which are used based on the needs of each application.

Programming languages: They must write the code to bring the application to life. Languages such as Python and R are widely used in developing AI applications thanks to their ease of use and the wide availability of libraries and packages specific to this task.

Big Data: AI requires much data to learn and improve. Big Data technologies allow for the storage, processing, and analysis of large amounts of information, which is crucial for the success of applications.

Cloud Computing: Some AI applications require a lot of computing power and computer resources, making it challenging to develop on a single computer. Cloud Computing allows access to many resources and computing power remotely, which is ideal for this type of application.

Data visualization: It is essential to be able to visualize the results of AI applications clearly and simply, and for this, data visualization tools are used. These tools allow for the graphical representation of information, making it easier to understand and allowing for the identification of patterns and trends.

Internet of Things (IoT): Integrating sensors and IoT devices in AI applications allows for collecting real-time data and its use to improve decision-making and optimize processes.

In conclusion, the development of AI applications in engineering requires a combination of technical and technological knowledge, as well as a rigorous methodology and a focus on the project's objectives. With these elements, it will be possible to create innovative solutions that improve processes and increase efficiency and effectiveness in engineering.

CONCLUSIONS AND RECOMMENDATIONS

AI is a technology that has rapidly evolved in recent decades and has significantly impacted a wide variety of industries, including engineering. AI in engineering improves current simulation and mathematical calculation methods, allowing for identifying patterns and unknown solutions.

Automatic learning systems, including linear regression, decision trees, and neural networks, are some of the most used tools in applied AI in engineering. These tools allow for the creation of simulation and mathematical calculation models with great accuracy and speed, improving the efficiency and effectiveness of engineering design and production processes.

However, implementing AI in engineering also poses some challenges, including the need for appropriate technological infrastructure, lack of knowledge and experience in using AI technologies, and the need to establish ethical and safe protocols for using these technologies.

In terms of developing AI applications in engineering, having an interdisciplinary team that includes AI experts, engineers, and mathematicians is crucial. In addition, specialized tools and technologies are required to implement AI in engineering projects efficiently.

7.1   Conclusions on the Application of AI in Engineering

AI is a crucial technology for the future of engineering, and its application continues to increase in different engineering fields. Therefore, engineers must be aware of the tools and technologies necessary for developing AI applications, as this will allow them to improve their skills and lead in adopting this technology. In conclusion, AI is a valuable tool for engineers, and its use will continue to grow in the future.

7.2   Recommendations for Implementing Artificial Intelligence in Engineering

Implementing AI in engineering is a complex process requiring considerable technical knowledge and programming skills. However, the effort invested in its implementation can result in significant findings and solutions that were previously unimaginable. Therefore, it is vital to consider the following recommendations:

1. Deep knowledge of AI: Engineers must have a deep understanding of AI, its history, types, applications, and challenges before implementing it in their daily work.

2. Clear identification of problems and opportunities: Before implementing AI, it is essential to clearly identify the issues and opportunities that can be solved or improved with its use.

3. Selection of appropriate tools and technologies: It is essential to select the proper tools and technologies for developing AI applications, considering the specific requirements of the project and the industry.

4. Acquisition of high-quality data: AI heavily relies on the quality of the data used for its training, so it is crucial to acquire and prepare high-quality data to improve the results of the application.

5. Continuous evaluation of results: It is essential to continuously evaluate the effects of AI implementation, to ensure that the desired improvement is being achieved and to make necessary adjustments and

improvements.

6. Collaboration with AI Experts: It is recommended to collaborate with AI experts to obtain a deeper perspective on the possible applications and solutions.

7. Inclusion of AI in Strategic Planning: The implementation of AI should be included in the company or project's strategic planning to ensure it effectively integrates into the overall processes and goals.

8. Training and Continuous Learning: Stay updated with the latest trends and advances in the field of AI and its application in engineering by participating in conferences, events, and continuous learning programs. This will allow you to be up-to-date with the new tools, techniques, and technologies available to improve the implementation of AI in engineering.

9. Data Privacy and Security Management: The implementation of AI should be done while considering the privacy and security of data, ensuring the protection of confidential and sensitive information. Additionally, it is essential to ensure that AI is used responsibly and ethically correctly.

10. Integration with other Systems and Technologies: AI should be integrated

with other systems and technologies in engineering to optimize its effectiveness and efficiency in problem-solving. Combining AI with other technologies, such as virtual reality, robotics, and the Internet of Things, can lead to greater decision-making effectiveness and solving complex problems.

Finally, it is essential to note that implementing AI in engineering requires a collaborative approach involving experts from both areas to ensure its success. Therefore, it is necessary to promote interaction between engineers, mathematicians, and AI experts to develop innovative and efficient solutions. In summary, implementing AI in engineering is an opportunity to improve processes and outcomes and achieve greater efficiency in the industry.

7.3   Future Perspectives of Artificial Intelligence in Engineering

AI is a technology that is constantly evolving and growing, and its application in engineering is an area that is in total development. Despite the current challenges and limitations, AI has shown its ability to improve simulation and mathematical calculation methods and discover unknown patterns and solutions. As technology advances, AI in engineering is expected to have an increasingly more significant impact on the world of engineering and technology in general.

In the future, AI will continue to be a valuable tool for engineering in various areas, including simulation and mathematical calculation, pattern detection, and problem-solving. Additionally, AI will be used to design and construct more advanced and efficient systems and structures and optimize production processes.

Another area where AI will have a significant impact in the future is automating engineering decision-making. AI can analyze large amounts of data and use advanced algorithms to make decisions based on previously unknown patterns and solutions. This will allow engineers to make more informed and efficient real-time decisions.

Also, AI will be used to improve safety in engineering. AI will continuously monitor systems and structures to identify potential risks and predict and prevent accidents. In this way, AI will play a key role in ensuring the safety of workers and the general public.

In conclusion, the future of AI in engineering is very promising, and its application will bring about numerous improvements and efficiencies in the field. Therefore, it is essential to continue researching and developing AI in engineering to take full advantage of its potential and to overcome its current challenges and limitations.

EPILOGUE

And so, with Artificial Intelligence revolutionizing engineering, we reach the end of this book. But it's not the end of the story, and it's just the beginning. AI has opened the doors to a new world of possibilities in engineering, where the limits are only those we set in our imagination.

In this book, we have explored how AI transforms how we plan, design, and build. We have seen how machine learning algorithms can analyze complex patterns and predict outcomes with unparalleled accuracy. And we have discovered how simulation and mathematical calculation models are evolving thanks to AI, allowing us to simulate and evaluate scenarios more efficiently and accurately.

However, AI also poses substantial challenges, such as data protection and privacy, as well as ethical implications regarding equity and diversity. We must address these challenges responsibly and consciously, ensuring that AI is used to improve all lives, not just to improve or benefit a few.

In conclusion, Artificial Intelligence in engineering is a revolution in the field, with a bright and limitless future. AI will continue to surprise us with innovative and impressive solutions as we continue to learn and develop new technologies. But this is just the beginning!